HEAT

HEAT

The Nature of Temperature
And Most Other Physics

Marc E. King

To order additional copies of this book, contact:
Xlibris Corporation
1-888-795-4274
www.Xlibris.com
Orders@Xlibris.com
131952

Contents

Forward by the Author 7

Introduction..9

Review And Background 11

    Proof of 3, 5, and 8 Dimensions 14

I. Proof of Dimensional Temperature—Water Example 1 17

II. Proof of Dimensional Temperature—Water Example 2 20

III. Proof of Dimensional Temperature—Nitrogen Gas Example 23

IV. Temperature of the Human Body 26

V. The Reason for Our Exact Body Temperature..........................31

VI. Entropy 34

VII. Mathematics, The Base of Natural Logarithms....................36

VIII. Short (Rydberg) Chemistry Review....................................40

IX. Tables of Dimensional Elements 44

    Table 1: Atomic E_{CR} 44

    Table 2: Nuclear Curvature C 45

    Table 3: Nuclear Intersections D = 3 46

    Table 3-1: Intersections for the Human Body 47

    Table 5: Nuclear Intersections D = 5 48

    Table for D= 2 49

X. Understanding Table 5 51

XI. Diagrams..56

    Diagram 1: Spatial Concept of 3 vs. 5 Dimensions 57

    Diagram 2: Spatial Concept of 5 vs. 8 Dimensions

    (*Cold Fusion*) 58

    Diagram 3: Summary (diagram form) of the text

    Fifth Dimension 59

Diagram 4: Working model of the Helium Nucleus 60
Diagram 5: Model of Closed Spatial Travel 61
Diagram 6: Hydrogen Fine Structure Progression
(Forward and Reverse, *Cold Fusion*) 62
XII. *Aside*: Dimensional Exercise: Memory vs. De Ja Vu..................63
XIII. Technical Summary..73
XIV. Reference Text and Appendices from Manuscript 75
XIV i Manuscript 77
XIV ii Appendix A 86
XIV iii Appendix B 88
XIV iv Appendix C 90
XIV v Appendix S...92
XIV vi Appendix X...97
XIV vii Appendix Z 99
XIV viii Summary of *Changing Your Mind* 104
XIV ix Summary of *Fifth Dimension* 107
XIV x Summary of *Cold Fusion* 108
XIV xi Summary of *The Nature of Mass* 110
XV. Reference 117
References...117
Reference Texts...118

Forward by the Author

Heat relates the prior three-dimensional concept of temperature to the present model of spatial progression.

Building upon facts shown in **The Nature of Mass**, the energies to and from higher dimensional space equate to the concept known as "temperature."

The text begins by using the model t=cB to exactly define the concept of time-based "temperature."

We start with the temperature (traversal energy) regarding and defining water (H_2O) on the planet surface. We next use nitrogen (N_2) as an example and as a proof. These serve to define phase-change purely from theory without the use of empirical results. The theory, in advance, predicts the correct empirical results.

Human body temperature is examined and determined by spatial theory alone. There is no laboratory involved. We do not need thermometers; instead, we begin this subject in several ways by proving empirical facts with the sequential spatial model alone as in the beginning of *The Nature of Mass*.

This text explains future nuclear technology, but also explains that nuclear technology is as simple and as natural as the blowing wind. Peaceful cold-fusion is the desired result.

The concept of "weapons" is not reviewed in this technical text.

Closed spatial traverse ("time" travel) is advanced in this text from the basis recorded in *The Nature of Mass*.

Introduction

As much as the author desires to have this text enjoyed and understood by everyone, there is much technical text required to show the concepts (which will be under technical scrutiny.) Denial is the first step in any problem solving or advancement process. Denial means "no, there is nothing more to understand." Without surpassing step 0, we can never achieve even step 1.

With all due respect, if the concept of continuous time were mathematically correct, there would never have been a need for quantum mechanics.

Now that we have seen quantum mechanics break down, we continue in denial and make up "attributes" like "color, etc." to explain-away what we do not understand. "Particles" are the most prolific concepts to salvage denial for the antiquated 3-dimensional "continuous time" concept.

There is a better way.

Please follow the math in this text.

Review And Background

We have postulated and subsequently proved the relationship[i(XIV)]

t=cB

and the resulting expression[viii(XIV)]

$$\mathbf{E/E_B = mV_B}$$

where:

t represents the perception of time in seconds

c is the numerical value of the "speed of light"

B is an integer, and

E_B has the units; energy mass^{-1} meter^{-3}.

This is a difficult and abstract concept because it removes time "t" as a unit of measure except for classical macroscopic physics.

One second of time t is replaced by c events B. In other words, time is not fully continuous and one second of time can be thought of as c (3E+08) discrete events (we generally use the Joule system of measure while we modify the system to alter the perception known as time t.)

In this model, c is no longer a velocity (because velocity requires the concept of continuous time t;) instead, c is a "number" of events and represents a constant "rate" of spatial progression (the progression of spatial frames.)

If B were a real and continuous variable, then the expression t=cB would simply represent a transformation of variables and would exactly and perfectly match all of existing physics.

But B is an integer, represents quantum (discrete) levels, denies all *continuity* and replaces it by slight **contiguity**, and changes non-macroscopic physics dramatically.

If time t were a real and continuous variable (it is not,) then the Schrödinger equation (a differential equation with respect to time t) would be exactly correct; however, time t is not real and is not continuous.

This is exactly the same concept as the Bernoulli[1(XV)] "compound interest" research that led to the definition of e, the base of natural logarithms.

To briefly state results from the manuscript *A Transformation of Variables Defining Space-Time and the Constant h*[i(XIV)].

$t = cB$
$E_B = 680eV \ kg^{-1} \ boundary^{-1}$
$b = b_3 = 1.111E\text{-}17 \ meter$
$E \ / \ E_B = \kappa c \ / \ n$
$\kappa = 0.548$

It is especially important to understand the concept of the energy ratio E_{CR} to get the most from the present text. There is an extended summary of *The Nature of Mass* attached as reference.

Several appendices from the original manuscript are attached to this text for additional reference.

This text is all about **PROOF**. *To end the Background Section, we define the Hydrogen Fine and Hyperfine energy structures from the spatial theory (t=cB) alone without* *need for any empirical results except to prove the exact correlation.*

*For readers who have followed the text **Cold Fusion**, you may have already performed this proof yourself.*

Fine and Hyperfine Energy Structures

Proof of 3, 5, and 8 Dimensions

The Hydrogen example of fine and hyperfine energy is illustrated in the diagram below:

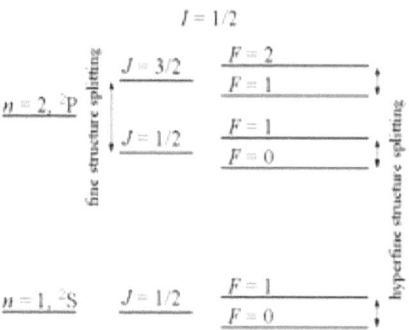

H in a hydrocarbon formation is normally in state 2. Per Diagram 6 (same as Cold Fusion Diagram 3,) there are exactly 2 opportunities or 2 occurrences of the exact state 2 energy within a single (1) frame of our three-dimensional experience.

We can determine energy from dimensional spatial theory, e.g. for the spin split energy 2_p denoted by J above:

$$\Delta E = F(R) \times \Delta R,$$

$$F = K m_p m_e / R^2, \text{ and}$$

$$E = K m_p m_e / (2^2 r_B)^2 \times (b_3/4) \times (10^3 \text{ meter}^3) \text{ where}$$

$K = 7.57E+28$ Nwt kg^{-2} m^2 (*Cold Fusion*)
$m_p = 1.67E-27$ kg (*The Nature of Mass*)
$m_e = 9.11E-31$ kg (*The Nature of Mass*)
$r_B = 5.29E-11$ meter (Bohr radius, Appendix Z) and
$b_3 = 1.111E-17$ meter.

Then E (spin-split) = 7.144E-24 J.

E = 7.144E-24 J / 1.602E-19 J/eV or

E (Fine) = 4.46E-05eV.

QED

Similarly, for the hyperfine energy denoted by F above,

$$E = K m_p m_e / (2^2 r_B)^3 \times b_5 / 4^3 \times (10^{(5-3)} \text{ meter}^2) \text{ where}$$

$b_5 = b_3 / c$ (*Changing Your Mind.*)

Then,

E (Hyperfine) = 4.396E-06eV.

We have not researched the accuracy of the low energy Lamb[7(XV)] experiments, and we have also averaged the radius for H state n=2 at $4r_B$.

I. Proof of Dimensional Temperature

Water Example 1
H_2O

To prove the dimensional application summarized in Table 3, let's use the dimensional elemental energies of traversal (our definition of temperature) to calculate properties of water at STP (standard temperature and pressure.)

As in the text "The Nature of Mass," we need no empirical results; instead, we independently calculate results using only the model t=cB and then show that the model exactly matches existing empirical results.

H_2O

H: $C_{5-8} = 0.0 \rightarrow +680eV/kg \times 2 = +1360eV/kg/boundary$
O: $C_{5-8} = 0.5 \rightarrow -1153eV/kg \times 1 = -1153eV/kg/boundary$
Total for molecule: $\qquad\qquad +207eV/kg/boundary$

Molecular curvature:

$C_{5-8} = Nn/(Nn+Np) = 8/18 = 0.444$
$C_{3-8} = 1 - C_{5-8} = 0.555$

Considering **1 mole** of H_2O, then

mass = 18amu x 6.022E+23 x 1.66E-27 kg amu^{-1} = **1.80E-02 kg.**

From "Review and Background"

$E_T = E_B \times mV_B$, where

m **(mole)** = 1.80E-02 kg
V_B **(STP)** = $4\pi r_E^2 \times b$
r_E = Mean Earth Radius = 6.371E+06 meter
$b = b_3 = 1.111E-17$ meter
E_B = 680eV/kg/boundary = 1.089E-16 J/kg/boundary.

Then, we calculate:
E_T = **1.11E-23 J or**
E_T = **6.94E-05eV.**

Summarized in Table 3, this energy-of-temperature represents a single event, a single transition, or a single boundary traversal involving 3, 5, and 8-dimensional space.

To transform E_T into existing 3-dimensional nomenclature for comparison, we must remove the 5-to-8 part and include only the 3-to-8 (three dimensional) part.

We also must adjust units to allow for the one-second of "time" t inherent in all existing measurement systems.

At STP, the 3-dimensional unit (proportionality) comparison volume is exactly 1 met³ and the 3-dimensional energy for *one mole* becomes:

$E_{T3} = 6.94E-05eV / C_{3-8} / 10^{-3} = 6.94E-02 / 0.555$ or
$E_{T3} = 3.86E-02eV.$

To complete the proof, this result must be equated to E_{T3} from the Boltzmann relation E(molar) = 3/2 kT where

k = 1.381E-23 (Joule system)
T = 299K (STP)

Then the Boltzmann[2(XV)] E_{T3}:

$E_{T3} = 3.86E-02eV.$

QED

II. Proof of Dimensional Temperature

Water Example 2
H_2O

Let's continue to calculate, solely from spatial theory, additional known natural facts of water, e.g. the density of liquid water at STP.

As shown in Example 1, H_2O loses 207eV kg^{-1} boundary^{-1} through the sequential spatial traversal.

To investigate density, we need to consider only the remaining energy for the equation $E / E_B = mV_B$.

Then $E / E_B = (680 - 207)/680 = 0.696 = mV_B$.

We now set $\rho' = m / V_B = 0.696 / V_B^2$ where V_B is defined in Example 1.

ρ' has units: mass x 10^{-3} / (met² boundary) from its calculation in example 1,

$(V_B = 4\pi r_E^2 \text{ x b.})$ Also, from $\rho' = .696 / V_B^2$,

$\rho' = 0.205$ molecule / boundary²
$\rho' / b = 1.845E+16$ molecule / boundary³
$\rho' / b / (6.022E+23) = $ mole / boundary³

mole-density $= 10^9$ x ρ' / b / 6.022E+23 / $C_{3-8} = 55.2$ **mole met⁻³**.

For 18 amu water, mass mole⁻¹ = 1.80E-02 kg, then

55.2 mole met⁻³ x 1.80E-02 kg mole⁻¹ x 10^3 = 994 kg met⁻³.

Then for no-neutron water, $\rho = 994$ kg met³.

QED

Exercise:

From the natural empirical knowledge that ²H is .0156% of all Earth-bound H, prove that the empirical density of water on the planet surface is 997 kg met³ at STP.

Hint: Use changes to molecular curvatures.

III. Proof of Dimensional Temperature

Nitrogen Gas Example
N_2

Let's calculate the number of moles in one cubic meter at STP. This result can easily be compared to classical time-based empirical physics.

Using only the spatial theory, we calculate as follows:

C_{3-8} = 1 - C_{5-8} = 0.5 for Nitrogen. C_{5-8} = Nn/(Nn+Np) = 7/14 = 0.5.

$E / E_B = mV_B = (-1153 / +680) \times 2 = 3.4$ (kg boundary) on the planet surface (see Tables 3 and 3-1.)

Set $\rho' = m / V_B$, then

$\rho' = 3.4 / V_B^2$ where

$V_B = 4\pi r_E^2 \times b$ (same as Water Exercise 1)

and we have acknowledged the (-) sign to mean energy "gain" through traversal.

Then,

$\rho' = 1.06E+05$ molecules boundary^{-2} and

$\rho' / b = (1.06E+05) / (1.11E-17) = 9.55E+21$ molecules boundary^{-3} or

$\rho' / b / 6.02E+23 = 1.6E-02$ mole boundary^{-3}.

For comparison, we must match this result to time-based units, similar to the water examples. Then, for STP:

"mole density" $= (1.6E-02) \times 10^3 / C_{3-8} = 16.0 / 0.5$ or

mole-density = 32.0 mole met^3 for N_2 gas.

From "time-based" macroscopic physics:

$\rho(N_2) = 0.895$ kg met^{-3} at STP.

For 1 mole,

mass(N_2) = (6.022E+23)(28 amu)(1.66E-27kg amu^{-1}) = 2.80E-02 kg.

Then 0.895 kg met^{-3} x (1 / 2.80E-02) mole kg^{-1} = 32.0 mole met^{-3} or

mole-density = 32.0 mole met^3 for N_2 gas.

QED

IV. Temperature of the Human Body

Table 3 *The Nature of Mass* has been modified and reproduced as Table 3-1 in this text to show the nature of "temperature." Table 3 is also attached in its original form.

There are many ways to equate the concept of temperature to the real sequential spatial universe. One of them is to advance the following Boltzmann idea[3(XV)],

E = 3kT/2 where

k = Boltzmann's constant = 1.381E-23 (J/K).

The concept of temperature is better explained in terms of mass and space.

Consider the human body temperature.

Unit human mass is made up as follows per Table 3-1:

Oxygen O 65%
Carbon C 18%
Hydrogen H 10%
Nitrogen N 3%
Calcium Ca 2%
Phosphorus P 1%

Total = 99% of mass.

Per Table 3 and Table 3-1:

$E = 3kT/2$ for the perceived 3-dimensions and using the concept of continuous "time," then

$969.1eV \ (1.602E\text{-}19 \ J/eV) = 3 \ k \ (J/K) \ T(K) \ / \ 2$ or

$T(K) = 7.48E\text{+}06.$

Except, as in Appendix C (attached,) we are again doing math in dimensions higher than 3.

T (degrees C) = e^{-1} (7.48E+06$^{3/8}$ - 273.15) - (1+φ) = 37.0 C, or

Human body temperature:

= 37.0 C

= 98.6 F.

This precisely agrees with thermometers of all types.

While external mass (Sun, Earth) affects our external temperature, our internal body temperature is largely unchanged while we are alive and well.

A More Transparent Way to Understand Healthy Body Temperature

First, we rewrite the expression above:

T(C) = (7.48E+06$^{3/8}$ - e^0 (273.15) - e^1 (1+φ)) / e^1.

See the section "Background Mathematics Review."

The constant k and the concept of temperature have been pre-defined assuming 3 "degrees of freedom" regarding 1-dimensional motion (3-dimensional space.)

We now understand that "matter," as we know it, is a three dimensional intersection between five and eight dimensional space. While the power 1/3 has been built-into the concept of temperature and also the constant k, there are in fact 8 degrees of freedom, not simply 3.

To use the familiar 3-dimensional constant k, we must raise the absolute temperature to the power 3/8 in order to replace the power 1/3 with 1/8.

Because the concept of temperature leads to e^{kT}, similar to the (rejected) theory of complete continuity e^{Rt}, the math is similar to appendices A and C, attached.

Alternately

We could also rework a momentum integral to account for higher dimensional space while altering units to replace the concept of time.

For 3-dimensions, we would then arrive at the expression:

$\varphi k^0 e^0 E^{3/8} - (\varphi + 1)k^{3/8}e^1 = T\, k^{3/8}\, e^1$ for T(K). Then

T(C) = T(K) - 273.15= 37.0, and

T(F) = T(C) x 9/5 + 32 = 98.6.

Unlike the degree-F scale, the degree-C scale is an absolute unit scale simply offset from the degree-K scale by a constant (273.15) based on the history regarding the "freezing" point of water (=0C.)

(Exercise: Physically, why is the "freezing" point of H_2O normally 273 K on the Earth surface? Hint: See Tables 3 and 3-1. Would the value change as a function of E_B? Answer: No. Why not? Hint: See Appendix E.)

An alternate and perhaps more intuitive way to understand the series:

$$T_3(K) = T_{3+5+8}(K) - T_{5+8}(K)$$

with subsequent conversion from K→C = (-273.15) and from D5→D3 = $(1+\varphi)$.

Note: $1+\varphi$ and $1+3/5$, in other words, $1+\varphi$ and $1+\varphi_{Local}$ where $\varphi_{Local} = 3/5$, achieve the same result to 0.1 degree in this case.

V. The Reason for Our Exact Body Temperature

Alternatively, we can also calculate:

37.0 absolute degrees:

κ^{-6} = exactly 37.0, per Appendix Z (attached.)

As proved in Appendix Z, the factor κ (0.5478) represents a dimensional change across any sequential (adjacent) dimensional boundary.

We assert the physical meaning:

From Appendix Z "Further Examining Spin and E = hv," we have calculated v for the Hydrogen state 2 spin-split energy, where

$$v = (b_3\kappa^6)^{-1} \text{ meter}^{-1}.$$

This represents all communication between 3- and 5-dimensional space, as shown in reference.

Rearranging the expression, we can write:

$$v = (b_3\kappa^6)^{-1} = b_3^{-1} / \kappa^6. \text{ Then,}$$

$$v = v_{MAX} / 37.0 \text{ where}$$

$$v_{MAX} = b_3^{-1} \text{ in three dimensions.}$$

For the "fastest" possible (or most efficient) communication, we would need to achieve $v_{MAX} = 37.0 \times v$ (for our H spin-split transition model.)

Alternately, the maximum achievable energy (hv) is proportional to the gap-width or "band-width" for any quantum energy state difference, including the H state 2 spin-split energy difference. Then,

$$v_{MAX} = 1/b_3.$$

While 37.0 is also a % difference between the entropic phase-change freezing and boiling points of H_2O, more importantly it is the constant κ^{-6} representing the factor of difference between adjacent dimensional "radial length" per Appendices A, C, E, Q, W, Z and also *The Nature of Mass*. In other words, it is the exact factor required to replace the 3-dimensional concept of temperature allowing for higher dimensional degrees-of-freedom.

We can then also write,

$$v = 1 / (b_3 \, T_{CB}) \text{ meter}^{-1} \text{ where}$$

$b_3 = 1.111\text{E-}17$ meter and
$T_{CB} = 37.0 = \Delta T(\text{abs factor}) = $ Body temp / 1-degree C. (K may also be used.)

In this case, starting with a single hydrocarbon kernel (suggested in Appendix X, attached,) we have evolved over 700 million years on an "intersection-surface" and have evolved our body chemistry specifically to maximize the thought process energy requirement using both 3 and 5 dimensional space and energies, i.e. we have adapted to the universe in which we and our minds have evolved and live.

VI. *Entropy*

Entropy is best acknowledged as "phase change of matter."

Phase change from solid to liquid to gaseous phase is simply defined by mass x volume per Appendix E.

For a given volume of space, mass x volume laws require minimum mass, i.e. the example elemental (intersection) energies from Table 3 must gain achievable translational energies (higher effective temperatures) to minimize mass x volume (minimize E_B.)

Temperature for a given element, including ambient conditions, or for a combination of elements, is not a statistical mystery; instead, it is the simple mass x volume law defined in Appendix E.

This is the reason water (H_2O) "evaporates" from liquid to gaseous form. Because the molecule has the distinction of "giving up" energy through the sums per Tables 3 and 3-1.

Consider the hydrogen (H) in "frozen" water (H_2O) known as ice, as on an ice hockey rink, as in an ice cube, or as a water mass at the North or South Pole.

Per Appendix Z (attached,) the H state-2 spin-split width (in meters) and the corresponding ($E=hv$) value for v (meter^{-1}) are the smallest possible for the element H ($1/37b_3$.)

As shown above, the largest possible $v = 1/b_3$.

As defined, the liquid form element is in a phase transition between a solid state and a gaseous state and in fact has become the basis of the Celsius temperature scale from 0 (frozen) to 100 (boiling) at standard atmospheric pressure.

In the water frozen state, the H spin-split energy can be called x. At 37 degrees Celsius, the same energy (spatial width) is exactly 37x. At 100 degrees Celsius, the energy remains 37x.

(Exercise: Per the math background section, what is the power (D) of the sequence that includes the integer 37? What is the sequential placement?)

VII. Mathematics

The Base of Natural Logarithms

In our perceived 3 dimensions (D=3):

(from 1 to e) $\int(1/x)dx = 1$.

In our actual 5 dimensions of thought (D=5):

(both from 1 to e) $\iint(1/xy)\partial x\partial y = x(1+1/x)$ where $y = 1/x$.

Sequentially, this is the sequence of positive integers.

The definition of "e" in fact means continuity and spatially defines real numbers.

Mathematics does not define space; instead, it is the other way around.

Similarly, the sequence of "area" for n integrals from 1 to e:

$$\{(\text{for } n) \int ..\int x^n(1+1/x^n)\partial \ldots n\} = \{n^n + 1\}.$$

In this text, we apply this sequence to the sub-sequence F(n=D) where:

F(D) is the Fibonacci infinite sequence per the attached Appendix C,

$$\{0,1: 1,2,3,5,8,13,21, 34, 55, \ldots \}.$$

Note: We have used nomenclature for the sequence F(n) such that F(n)=D, e.g. F(n=5) = 3 → D=3, F(n=6) = 5 → D=5, etc.

Important Infinite Sequences:

As we have transcribed,

$\varphi = \lim(n\rightarrow\text{infinity})$ F(n-1) / F(n) as the Fibonacci sequence converges,

$\gamma = \lim(n\rightarrow\text{infinity})\ F(n-2)\ /\ F(n)$ as the Fibonacci sequence converges,

$e = \lim(n\rightarrow\text{infinity})\ (1 + 1/n)^n$ as the Bernoulli sequence converges.

As we have shown in Appendix C (attached) and also (**Fifth Dimension**) Appendix S (attached,) the dimensional rate of increase for the base of natural logarithms:

$\{R_{LogBase}\} = \{(1/\gamma)^{3/1}, (1/\gamma)^{5/2}, (1/\gamma)^{8/3}, (1/\gamma)^{13/5}, \ldots\}$ and

$\lim (D\rightarrow\text{infinity})\ R_D = (\gamma^\wedge-1)^{(\gamma^\wedge-1)} = 12.42$ where

$\gamma = 1 - \varphi.$

Then,

$e_{(D)} = e_{(D+n)}\ \text{x}\ 12.42^{-n}$

where we have used similar nomenclature such that

$e_3 = e_5\ /\ 12.42^1$
$e_5 = e_8\ /\ 12.42^2$

and so on.

We have now calculated our additional decimals for the factor 1.08 derived in the Manuscript and Appendices A, B, C, and S.

1 / 12.42 = .0805 as the infinite sequence $\{R_{LogBase}\}$ converges.

VIII. *Short Chemistry Review*

Rydberg[4(XV)] *Molecule*

A Rydberg molecule is distinguished by its large size (volume.)

It is quite unstable due to its mass x volume, or we could otherwise say the "far-away" electron isn't bound very "tightly."

Either way, it is difficult to "make" and doesn't last very long[5(XV)].

Consider the potentially useful Rydberg molecule:

$NH(CH_3)_3$.

At once, this molecule possesses everything of interest:

1. Approaches mass x volume 3-dimensional limits.
2. Has an H bonded to an N, a lower energy bond than a CH bond.
3. Includes several kernels CH_3.
4. Has the attribute "strange" meaning quantum physics cannot explain it.
5. At a given temperature, would "break" its NH bond before breaking an CH energy bond and thereby "spend" energy required for the $E_B = 680eV$ that would otherwise serve to reduce necessary hydrocarbons.

Unfortunately, due to the shear size of the molecule[6(XV)], it may be toxic and unable for use to reduce the rate of hydrocarbon decay in the body.

Consider the spatial (dimensional) transition energy per kg for the molecule:

C x 3 = -3,459 eV
H x 10 = 6,800 eV
N x 1 = -1153 eV

Total weighted molecular energy requirement, as in Table 3-1:

E = 2,188 eV.

E = +2,188 eV (+ nomenclature meaning the molecule is required to lose the energy through a single transition and per unit mass, using kg as the mass unit to adjust magnitude as throughout the text and references.)

While this molecule has many desirable qualities, it should not "last" for many frames and is very large for the human body; however, in a given frame $(b_{3,})$ it is fully capable of "breaking itself up" instead of an established CH_3 (brain) or even an NH (protein) binding as shown in Appendix E.

In an alternate approach, a molecule such as "2-butene-1-thiol" or better "2-quinolinemethanthiol" could perhaps help to "act" as an "OH-similar" agent while remaining closer to the human hydrocarbon attribute form.

The molecular study is beyond the scope of this text, but the mass-volume principles remain clear.

Alternatively, we can look for different ways to accomplish the 3.4eV addition requirement per the health reference text[Text2(XV)]*:*

Add 3.4eV (visible wavelength) to the outer skin?

No one wants to live in a violet light-chamber, but the accomplishment is easy in the "reverse" direction by simply and efficiently "moisturizing" the skin and thereby providing H_2O to lose its 207eV / boundary instead of losing the same energy from our skin tissue (as per Section I.)

Add 3.4eV interior from the skin? This is not so easy.

There is already N and H in our blood streams (consider the energies from Section III - N_2 example.)

A catalyst(s) to provide the safe release of 3.4eV / n, where n is an integer, can accomplish this task, similar to the good but imperfect method of vitamin ingestion.

In terms of curvature and temperature:

The elemental and/or molecular curvature $\frac{1}{2} = 0.5$, e.g. He, N, N_2, O and O_2, provides the energy-of-temperature of exactly $1153 / 680 = 1.7$ eV kg^{-1} boundary^{-1} on the planet surface as shown throughout the text and in Tables 3 and 3-1.

In other words, as examples, un-reacted N_2, un-reacted O_2, and He (almost always un-reacted) provide the 1.7 (x 2) = 3.4eV per unit mass per boundary to negate the enabling energy (-3.4eV) from completing the aging reaction of $-3.4 - \underline{10.2} = -13.6$eV.

Exercise: (a) What is the atomic curvature of N? (b) The molecular curvature of N_2? (c) The atomic curvatures of standard H and O? (d) What is the group curvature of OH? (e) Does the curvature C_{5-8} change for an hydroxyl group (OH) depending on electron presence?

Answer part e) - no.

IX. Tables

Table 1—Atomic Examples of the Ratio E_{CR}

Symbol	Nomenclature	Nn/Np	E_{CR} / 1833
H	Standard Hydrogen	0/1	0
^{2}H	Heavy Hydrogen	1/1	1
He	Standard Helium	2/2	1
^{55}Fe	Unstable Light Iron	29/26	1.12
^{56}Fe	Iron	30/26	1.15
^{57}Fe	Iron	31/26	1.19
^{58}Fe	Iron	32/26	1.23
^{59}Fe	Unstable Heavy Iron	33/26	1.27
^{235}U	Uranium	143/92	1.55
^{238}U	Uranium	146/92	1.59

Table 2—Nuclear Examples of Curvature C

Element	Type	AMU	Nn/Np	C Nn/(Nn+Np)
H	standard	1	0	0.000
H	heavy	2	1	0.500
He	standard	4	1	0.500
Fe	unstable	55	1.12	0.527
Fe	stable	56	1.15	0.536
Fe	stable	57	1.19	0.544
Fe	stable	58	1.23	0.552
Fe	unstable	59	1.27	0.559
U	uncommon	235	1.55	0.609
U	common	238	1.59	0.613
Pu	unnatural	244	1.60	0.615
Cm	unnatural	250	1.60	0.616
	unnatural			0.618

Table 3
Nuclear Intersections

Element Nucleus	Type	AMU	Ecr/1833 = Nn/Np	C_{5-8} = Nn/(Nn+Np)	C_{3-8} (1 - C_{5-8})	C Int = C_{3-8} + C_{5-8}	Inter-section	D (Int)	Boundary Energy Req. Earth (eV/kg)	Comment
H	standard	1	0	0.000	1.000	1.000	3-int-5	2	680.0	No 3-D Intersection
H	heavy	2	1	0.500	0.500	1.000	5-int-8	3	-1153.0	
He	standard	4	1	0.500	0.500	1.000	5-int-8	3	-1153.0	
Fe	unstable	55	1.12	0.527	0.473	1.000	5-int-8	3	-1373.0	
Fe	stable	56	1.15	0.536	0.464	1.000	5-int-8	3	-1428.0	
Fe	stable	57	1.19	0.544	0.456	1.000	5-int-8	3	-1501.3	
Fe	stable	58	1.23	0.552	0.448	1.000	5-int-8	3	-1574.6	
Fe	unstable	59	1.27	0.559	0.441	1.000	5-int-8	3	-1647.9	
U	uncommon	235	1.55	0.609	0.391	1.000	5-int-8	3	-2161.2	
U	common	238	1.59	0.613	0.387	1.000	5-int-8	3	-2234.5	
Pu	unnatural	244	1.60	0.615	0.385	1.000	5-int-8	3	-2252.8	
Cm	unnatural	250	1.60	0.616	0.384	1.000	5-int-8	3	-2252.8	
				φ 0.618	γ 0.382	1.000	NA	5	NA	No 3-D Intersection

Table 3-1
Human Body Temperature Energies

element name	element symbol	human %	Nn/Np	Ecr (x 1833)	Earth boundary energy eV/kg	boundary weighted avg
oxygen	O	65	1.0	1833.0	-1153.0	-757.0
carbon	C	18	1.0	1833.0	-1153.0	-209.6
hydrogen	H	10	0.0	0.0	680.0	68.7
nitrogen	N	3	1.0	1833.0	-1153.0	-34.9
calcium	Ca	2	1.0	1833.0	-1153.0	-23.3
phosph	P	1	1.1	1955.2	-1275.2	-12.9
tot		99				-969.1

Table 5
Nuclear Intersections
Example Table of 5-Dimensional Natural Elements

Element Nucleus	ISO	Nm13	Nn	Ecr EC_{13}/EC_8	$C_{8\text{-}13}$ $N_{13}/(Nn{+}N_{13})$	$C_{5\text{-}13}$ $(1-C_{8\text{-}13})$	C Int = $C_{5\text{-}13}+C_{8\text{-}13}$	Inter-section	D (Int)	Boundary Energy Req. Earth (eV/kg)	Comment
1H	3D-heavy	0	1	0	0.000	1.000	1.000	3-int-5	3	680.0	No 5-D Intersection
1	1	1	1	1	0.500	0.500	1.000	8-int-13	5	0.0	
1	2	1	2	1	0.333	0.667	1.000	None	X		No 5-D Intersection
1	3	1	3	1	0.250	0.750	1.000	None	X		No 5-D Intersection
2	1	2	1	1	0.667	0.333	1.000	None	X		No 5-D Intersection
2	2	2	2	1	0.500	0.500	1.000	8-int-13	5	0.0	
2	3	2	3	1	0.400	0.600	1.000	None	X		No 5-D Intersection
2	4	2	4	1	0.333	0.667	1.000	None	X		No 5-D Intersection
3	1	3	1	1	0.536	0.464	1.000	8-int-13	5	0.0	
3	2	3	2	1	0.600	0.400	1.000	8-int-13	5	0.0	
4	1	4	1	1	0.800	0.200	1.000	None	X		No 5-D Intersection
4	2	4	2	1	0.667	0.333	1.000	None	X		No 5-D Intersection
4	3	4	3	1	0.571	0.429	1.000	8-int-13	5	0.0	
5	1	5	1	1	0.833	0.167	1.000	None	X		No 5-D Intersection
5	2	5	2	1	0.714	0.286	1.000	None	X		No 5-D Intersection
5	3	5	3	1	0.625	0.375	1.000	None	X		No 5-D Intersection
5	4	5	5	1	0.556	0.444	1.000	8-int-13	5	0.0	
					φ 0.618	γ 0.382	1.000	NA	8	NA	No 5-D Intersection

TABLE for D= 2

A table is no use for the case D=2 because the results are straight-forward.

Similar to the derivation for E_{CR} in three dimensions, we can derive E_{CR} for the case of two dimensions traversing through three dimensions.

$E_{CR} = m_E / m_{-P} = (r_E / r_P) \times (1 + .08^{3/2}) / \varphi$ where

$r_E = 6.89E\text{-}18$ meter, and

$r_P = 7.68E\text{-}15 \times (1 + .08^{5/3})$ meter

per *The Nature of Mass* p. 11-12. (See Reference Texts.)

Then,

$E_{CR} = m_E / m_{-P} = 1.485E\text{-}03$ or

$m_{-P} / m_E = 673.4$ and

$m_{-P} = 6.134E\text{-}28$ kg.

Then,

$E_{CR} = m_P / m_{-P} = 2.723$ or

$E_{CR} = e \, (1 + .08^{2/1})$ where

e = the *base of natural logarithms* in three dimensions = 2.718

(within an error factor of .18% or .0018.)

E_{B2} is relatively large (~10^7 eV/kg) per the reference text *Changing Your Mind, A Mathematical Transformation of Variables Regarding Space-Time and the Constant h,* Appendix B.

Then E_{CR} in the case D=2 is a constant. There is no "chemistry" for D=2.

Table 2 (for the case D=2) then reads as follows:

$$C_\varphi = m_P / (m_P + m_{-P}) = 0.7315$$

$$C_\gamma = 1 - C_\varphi = 0.2685$$

$$C_I = 1.$$

There are no "elements" in two dimensions.

X. Understanding Table 5

Higher φ_{Int} (or higher intersection elemental curvature C) implies higher energy, i.e. higher E_{CR} and higher boundary (negative or gained) energy "supplied." Dimensions to not "stop" at 8.

1. Higher energy equates to additional mass.
2. Neutrons are the 8-dimensional mass form.
3. Any mass with the correct quantum value(s) should/ could occur.
4. The "extra" transitional (dimensional, low thermal) energy should add to "temperature" related to $e^{\wedge}kT$ in 3-dimensional space.
5. *In the current model, there is no real time t. Many things, including the concept of temperature, must be redefined*

6. *The Concept E=hv has been redefined in Appendix Z (attached.)*
7. *We can no longer relate or confuse the concept of energy with the concept of "repetition per second."*
8. *Per Appendix Z, "v" has the units (meter⁻¹.) There is no real time t.*

Dimensionally, we have shown that the ratio of closed-to-open is $22/7 = \pi$ in three dimensions. We have shown the following for higher dimensions:

$$3D \rightarrow \pi$$
$$5D \rightarrow \pi^3$$
$$8D \rightarrow \pi^6$$

These relate to the elemental values for energy E required for a single boundary traversal for one unit mass, e.g. one kg.

Exact ratios of "closed distance" / "open distance" or C(1) / C(0):

$^{235}U = (2161.2 / 680.0) / \pi = 1.011$
$^{238}U = (2234.5 / 680.0) / \pi = 1.046$
$^{244}Pu = (2252.8 / 680.0) / \pi = 1.054$
$^{250}Cm = ^{244}Pu = 1.054$

As we have shown (**Cold Fusion**,) once π has been exceeded, the next dimensional "curvature" is π^3.

The 3-dimensional natural curvature $C = \varphi = 0.618$ where $0 < C < 1$.

With our three dimensional relative curvature-of-closure ratio $= \pi$ as shown above, elemental spatial intersections (mass) having curvature $C_3 > \pi$ are no longer "curving outward." They have in fact surpassed closure and are spiraling inward through dimension D=5 as shown in *Fifth Dimension*.

We assert the physical meaning:

1. The 3-dimensional curvature φ and the 5-dimensional curvature φ are within a spatial intersection (mass) where the intersection curvature is exactly and always 1 per Table 3.
2. The intersection (mass) curvatures φ_I and γ_I are en-route, but have not yet arrived, at the five dimensional curvature $= 1$ having the new value π^3 in the same way we view the three-dimensional closed curvature to have the ratio π.

Consider the alternate view of the table above that stems from Table 3:

$E_{REQ} / E_B = E(\text{left over}) / E_B = E(\text{extra}) / E_B = E(\text{contributing to temperature}) / E_B = $ as follows:

$Pu = 3.313$

$$^{238}U = 3.286$$

$$^{235}U = 3.178$$

All less massive elements (intersections of dimensional mass) have the value:

$$E_{REQ} / E_B < 3.1429 \text{ or } < \pi.$$

In terms of closure C:

The naturally occurring element U, and the unnatural elements created from nuclear fission (Pu and Cm,) are close enough to the boundary $\varphi = .618$ (from Table 3) that they have difficulty to "remain" in 3-dimensional space.

In order to maintain a 3-dimensional existence, the "high-curvature" and relatively "heavy" elements must release energy, i.e. mass.

Consider the values of n (as in Appendix E) for these elements:

$$n_{PU} = \kappa c / 3.313 = 0.302 \, \kappa c$$

$$n_{238U} = \kappa c / 3.286 = 0.304 \, \kappa c$$

$$n_{235U} = \kappa c / 3.178 = 0.315 \, \kappa c$$

where n = κc represents $E/E_B = 1$, i.e. the transition from "bound = 3-dimensioanl mass" and "unbound = no 3-dimensional intersection."

And where κ = 0.5478,

$E = h\nu = hc / \lambda$,

$n = \lambda / b_3$,

and $b_3 = 1.111E{-}17$ meter.

We have already shown that ("closed distance" / "open distance") / $\pi = 1$ for closed 3-dimensional space (above.)

We have already shown that ("closed distance" / "open distance") / $\pi^3 = 1$ for closed 5-dimensional space (*Cold Fusion* and Appendix Z in the section "Closed-Spatial-Travel." See references.

XI. Diagrams

3 and 5 Dimensions

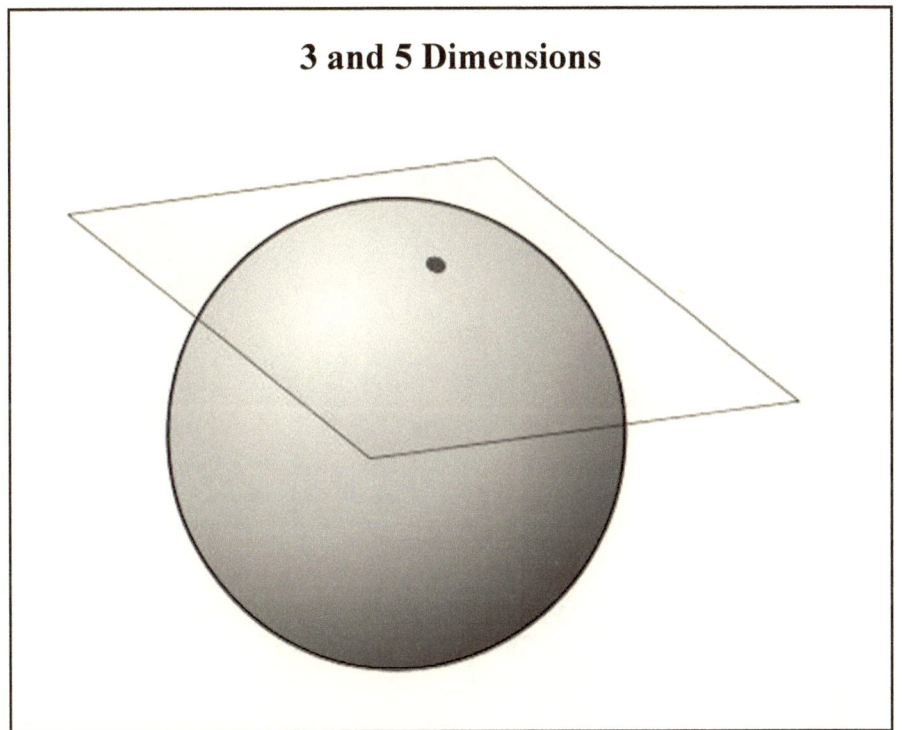

Coordinates for the 3-Dimensional Sequence

Three Dimensions Moving about Two Dimensions in Five Dimensional Space

Imagine the spheres progressing along a curved spiral arc while the extra-two-dimensional arrows are straight.

PREVIOUS
(xc,yc,zr,tx-1b,ty-1b)

CURRENT
(xc,yc,zr,0,0)

NEXT
(xc,yc,zr,tx+1b,ty+1b)

Figurative Model for Spatial Dimensions 3 + 2 = 5 and 5 + 3 = 8

3-Dimensional Intersection Between 5- and
8- Dimensional Space. A "Black Hole."

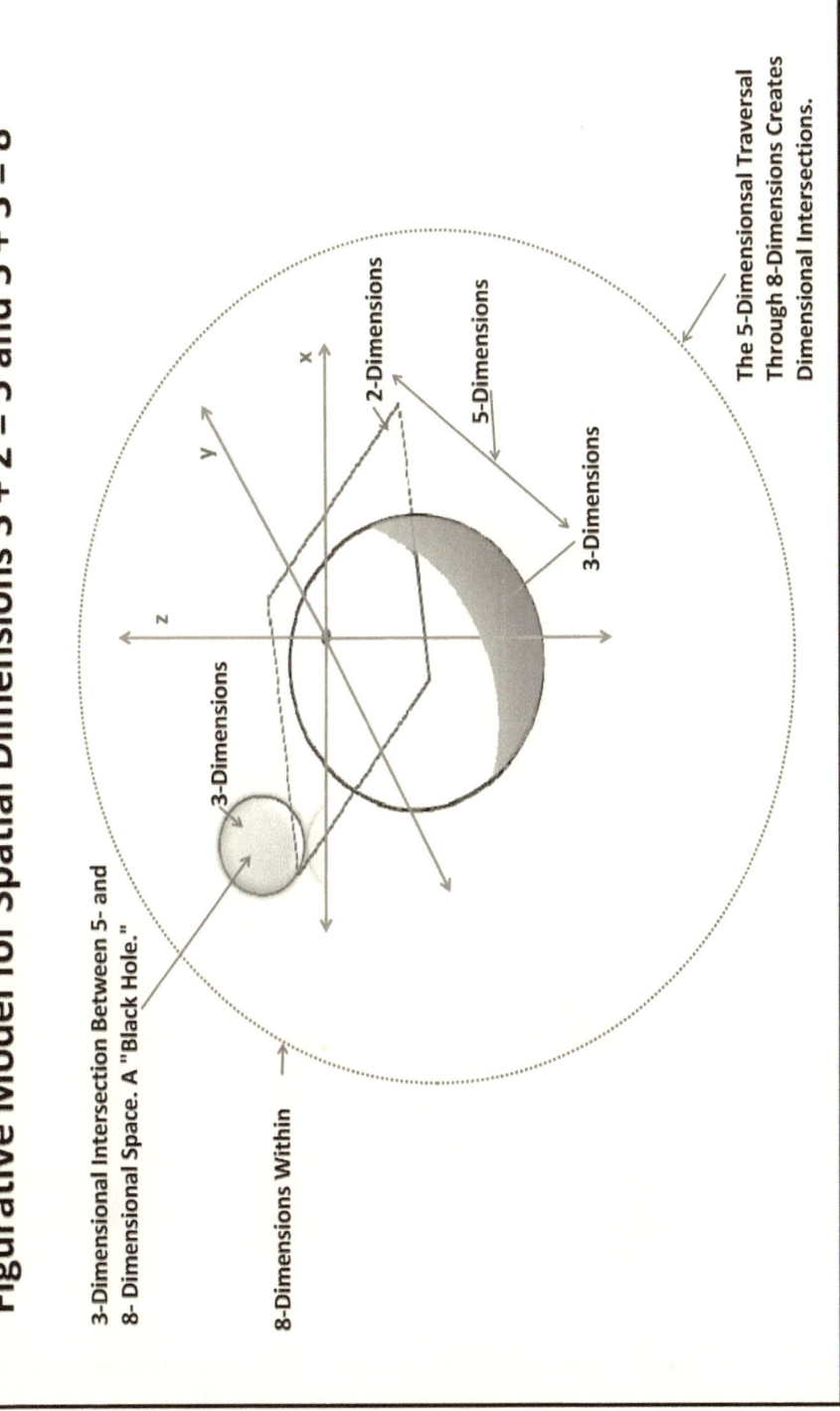

8-Dimensions Within

3-Dimensions

z

x

y

2-Dimensions

5-Dimensions

3-Dimensions

The 5-Dimensionsal Traversal
Through 8-Dimensions Creates
Dimensional Intersections.

Working Model of The Helium Nucleus

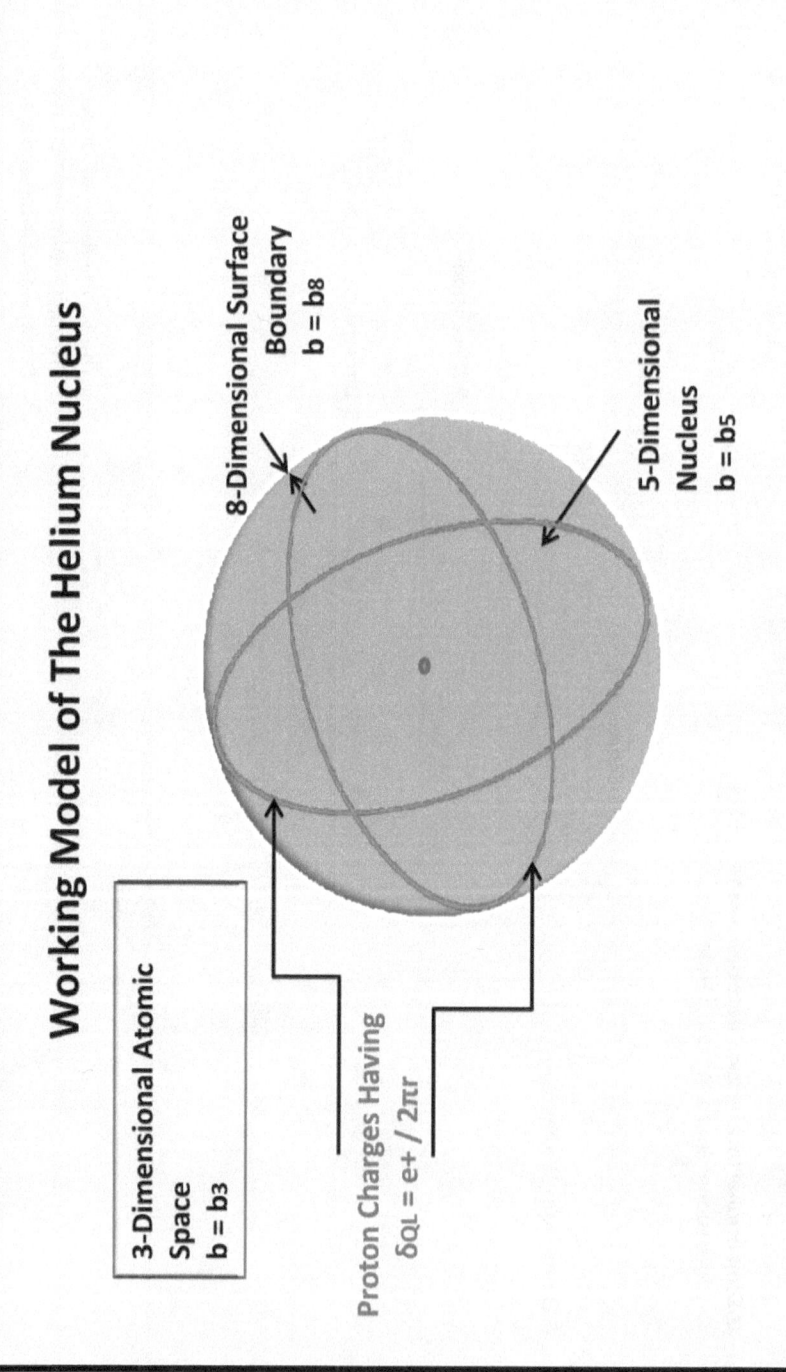

3-Dimensional Atomic
Space
$b = b_3$

8-Dimensional Surface
Boundary
$b = b_8$

5-Dimensional
Nucleus
$b = b_5$

Proton Charges Having
$\delta_{QL} = e^+ / 2\pi r$

The Charge Pair has $(1/b_5)^{\wedge}3$ Allowed Spin Positions

Fibonacci Spiral having Curvature C = φ

Showing Radius of Closure R$_C$

Not to Scale

R_{C5}

(x,y,z,0,+R$_{C5}$)
Example Coords.

Example Coords.
(x,y,z,0,-R$_{C5}$)

Forward and Reverse-Photons
Reading from Reverse and Writing to Forward

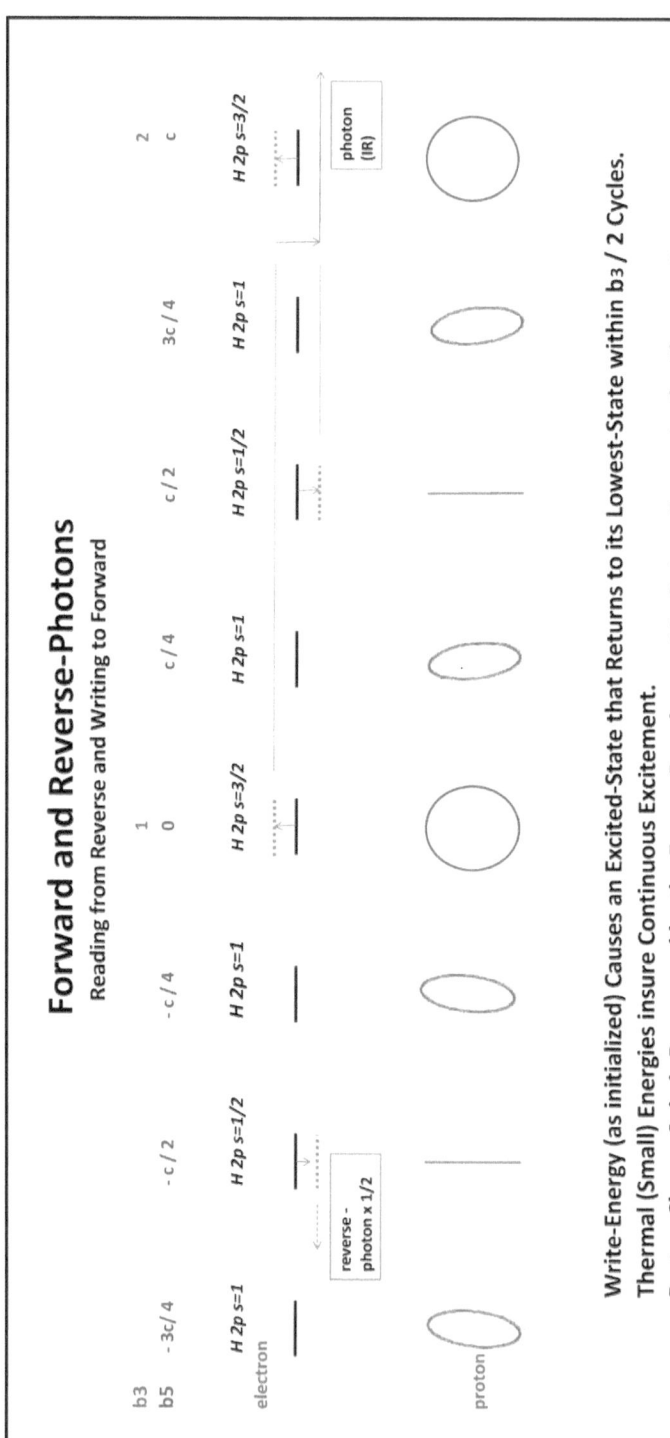

Write-Energy (as initialized) Causes an Excited-State that Returns to its Lowest-State within b₃ / 2 Cycles.

Thermal (Small) Energies insure Continuous Excitement.

Proton Charge Spin is Dampened by the Energy Requirement to Return Energy to the Electron States

Charge Motion Propogates in Both Directions, i.e. Forward and Backward.

XII. Dimensional Thought Exercise

Aside . . . Dimensional Intuition: Memory vs. De Ja Vu

A memory represents the traverse across 5 dimensional space to a pre-sequent 3 dimensional space and appears to our minds as 2 dimensional (like a photograph.)

De Ja Vu represents the traverse across 8 dimensional space to a pre-sequent 5 dimensional space and appears to our minds as 3 dimensional (like a local *sequence* of frames or a "short time of prior experience.")

As in 5 dimensions with respect to 3 dimensions, the 5 dimensional "closure" in 8 dimensions represents "running into the back of yourself" known in three dimensions as De Ja Vu. Instead of a photo image that we can visualize, the higher dimension allows us to "experience" as opposed to simply "visualize."

While memory is "at your command," De Ja Vu is not. In fact, you are "at the command" of your own De Ja Vu.

It is well known than people facing imminent death and who have survived recall their entire lives "flashing" through their minds. The "flushing" of memory addresses seems almost continuous. While it seems a continuous recollection of experience, it is not fully continuous.

De Ja Vu is an experience in 3 dimensions. Instead of "seeing" a photo type 2-dimensional image, you in fact "feel" in 3-dimensions that while you are there you have "been there before" in that exact same point in space and "time."

The reason, from the prior texts, is that your mind lives and exists in both three and five dimensional space; however, it does not understand eight dimensional space.

While your 5-dimensional mind can easily recollect past 3-dimensional frames, and can easily use the low-energy capability of 5-dimensional space in order to think, reason, imagine, and create, your 5-dimensional mind does not comprehend 8-dimensions.

As we determine the "addresses" of pre-sequent frames in order to remember, alternately, the higher dimensional addresses present themselves to us as a "surprise."

While memory is vested in the "past" or pre-sequent events, De Ja Vu has an indistinct "temporal" or sequential value so far as we know in our three dimensional world.

Physically speaking, to further understand, we should need to explore the only eight dimensional concepts we know, i.e. neutron mass on the exterior surface of atomic nuclei and on the interior surface surrounding the center of black holes.

It would be useful to determine a vector map from 5 to 8 dimensional space as we have determined the vector map from 3 to 5 dimensional space (Appendix Q, *Fifth Dimension*.) Fortunately, we are now able to extrapolate all vector maps from Appendix W, *Cold Fusion*.

Physics for the Two Concepts

To begin, we determine 't' the "time length of experience" for a memory and separately for a "De Ja Vu."

A Memory:

A single memory is a single frame (in two dimensions) of our physical recollection of an addressed spatial frame-width through 5 dimensional space.

Because the 3-dimensional width b_3 = 1.111E-17 meter, the effective "time-length" for the recollection is t for a single event B. Numerically,

$$t_3 = 1 / c = 3.333E\text{-}09 \text{ second.}$$

A "De Ja Vu:"

In 5 through 8 dimensions, there are c events per a single 3 dimensional event B (Appendix L, *Changing Your Mind*.)

Then $t_5 = t_3 \times c$ or, at a minimum,

$$t_5 = 1.0 \text{ second.}$$

In fact, the perceived "time length" of a De Ja Vu is altered by the fact of human synaptic speed of biochemical perception.

Anyone who recalls a significantly different time for a De Ja Vu experience may be recollecting other "time" involved with the perception process.

An Independent Determination
of "Length of Time"

Per the attached Appendix Z, ***Cold Fusion***:

$r_{c5} = cz^2b_3 / 2\pi^3$, then

"Equating" distance with the concept of time:

$cb_3 / 2\pi^3 = 1$ sec, and

t (sec) $= cz^2b_3/2\pi^3$ (represents the distance $cz^2b_3/2\pi^3$ 'meters' of spatial progression through 5-dimensions.)

Then the time length from "request" to "experience" for a De Ja Vu:

$t \leq -1.0$ sec.

We can explore ***one means*** of estimation (as follows) by using the ground-state simple value z:

Since our hydrocarbon chemistry is predominantly Hydrogen (z=1) and Carbon (z=6,) then our "differential" time from "request" to "De Ja Vu" should fall between -1.0 and -36.0 seconds.

Arguably, within one to thirty-six seconds from experiencing a De Ja Vu, you may expect to have an experience that causes your mind to "request" the review (De Ja Vu.) This means you should experience an intersection to your prior sequential frame, and that your "time" in between request and receipt (actual to intersection) will be as follows:

$-1(\text{sec}) \geq t \geq -36 (\text{sec})$.

In other words, it should take 1 to 36 seconds following a De Ja Vu for your mind to experience something (possibly anxiety or otherwise emotionally related) that will cause your mind to request the De Ja Vu.

We suggest a possibly **better means** of estimation below:

Per Appendix Z, we can also write,

$r_{c5} = cn\lambda b_3 / 2\pi^3$.

For Hydrogen (H) within a kernel -CH_3- where

$z=1$ and the state

$n=2$, then

$r_{c5} = cn^2 b_3 / 2\pi^3$.

For this case, we are physically going nowhere; instead, our minds have addressed the "location" of scrutiny after-the-fact or "subsequent" to a sequence of pre-sequent frames-to-be-reviewed.

In this case, as in the above calculation,

$t \leq -4.0$ second.

How to Request a De Ja Vu?

You should not be able to consciously request a De Ja Vu; instead, they seem to be a combination of subconscious information your mind possesses and internal subconscious thought that "warns" you "ahead of time."

A De Ja Vu is most likely not a sign of an impending problem, it is simply a subconscious indication that you may be uncomfortable with something or simply a sign that you have "perceived" a possible situation you may wish to avoid. It may mean nothing at all except that your mind is operating at a high level.

The physics of the communication can be intuitive, or the reader can follow the technical summary below.

A Physical Theory of De Ja Vu

We already know the mind has tremendous physical powers (*Cold Fusion, Dignity of Mind.*)

We already know that one second of our perceived "time" represents c (3E+08 in metric units) or 300,000,000 physical events.

We already know from the proved "five-dimensional-computer" (*Fifth Dimension, The Light to See*) that intellect (calculation, creation, operation, algorithmic result, thought, dreams, and much more) takes place "in-between" our 3-dimensional intervals of experience.

Our minds are capable of "calculating" our next "moment" of experience in-between the previous moment and the "present" moment.

In other words, each moment would seem as though we had "already experienced that moment."

But the forward calculation (premonition) of experience can last only as long as the "data" or sensory information clearly leads to the next step. For example, if you have a De Ja Vu experience as you are about to open a closed door, your De Ja Vu experience will most likely end immediately since your direct senses probably have no information beyond the closed door upon which to continuously build.

XIII. *Technical Summary*

1. We have independently derived, from dimensional spatial theory alone, without need for any empirical results, the definition of temperature.
2. We have used the derived facts to prove existing empirically based equations to three decimal points.
3. We have similarly done so for the concepts of mass density and molar density using ***mass x volume*** rules defined by the dimensional spatial theory.
4. We have derived the existing, verified mean temperature of the human body, along with the verification that 37C (98.6F) maximizes the communication "frequency" $v = v_{MAX} / 37$ for the H state-2 spin-split energy.
5. We have furthered the spatial curvature model that defines the periodic chart of elements as in ***The Nature of Mass***.

6. We have developed a molecular spatial curvature model.
7. We have developed (Table 5) the concept of dimensional mass (spatial intersection) to include an even higher dimensional model.
8. From spatial theory alone, we have shown and explained the exact fine and hyperfine dimensional energy splits observed for Hydrogen.

XIV. Reference Text and Appendices from Manuscript

Special Note Regarding References:

This section contains original text transcribed from the main manuscript, from appendices, and also from summary text contained in other works. These reference texts have their own sited references, including diagrams and figures, noted within each text.

Unless a special note is inserted, there is no attempt to include the sited references within each respective reference text, although each text may indicate one or more references.

References listed below in the section "XV References" are the sited references from the present text and only from the present text. Similarly, diagrams and tables within the present text are sited by the present text alone and do not necessarily correspond to diagrams and tables that may have been noted in any of the "reference" texts.

XIV i.

A Mathematical Transformation of Variables Defining Space-Time and the Constant h

Marc E. King
Silicon Valley, California
June, 2012

Abstract

A variable transformation for time t is supported by wave mechanics and relativity theory and shows that time and space can be related and connected by the concept of physical events per unit space. The transformation confirms our daily macroscopic experience as unchanged from classical physics while still suggests new physics regarding small energies and large spaces. Perceived time can be altered relative to Earth-bound clocks in regions of lower or higher gravitational force. A series of calculation-verifications proves the theory, derives Planck's constant and defines quantum mechanics. Black holes and their mass-radius relationship are defined. The Schwarzschild radius is defined. Minimum and maximum energies are defined.

Introduction

In this model, it is shown that continuous time t and a contiguous view of spatial frames are mathematically the same in the macroscopic sense. A suggested transformation of variables presents interesting differences in concept for small and large energies and spaces.

Concept

We postulate that continuous time as experienced can also be represented, with the same physical result, as a directional spatial sequence or frames of events.

We consider a new unit system using the transformation

$$t = cB$$

with c = "speed" of light, where one spatial frame (size b) is related to one physical event B by

$$b \text{ (meter)} = 1 \text{ (event)} / B \text{ (events meter}^{-1}.)$$

The transformation $t = cB$ implies the units t (sec) = c (met/sec) x B (sec^2 met^-1)

Then B events per meter = B sec^2 per meter,

and one physical event = one square second = sec^2.

Derivation

From wave mechanics, we have the Schrodinger equation[i]

$d\psi/dt = +/- 2\pi i/h \times E\psi$ as a partial derivative

and the related approximation

$\Delta x \Delta k \geq O(1)$.

This defines the uncertainty in measurements[ii]

$\Delta x \, \Delta p \geq h/2\pi$.

Implying $\Delta p = m \, \Delta x / \Delta t$ and using a transformation for Δt, then

$\Delta p = m \, \Delta x / \Delta(cB)$

This leads to

$m \, (\Delta x)^2 \geq (h/2\pi \text{ J-s}) (\Delta cB) = (h/2\pi \text{ J}) (cB) (\Delta cB)$

Per unit mass, then

$(\Delta x)^2 \geq (h/2\pi)(\Delta cB)(cB)$ from transformation.

For a single B (events-meter^-1) the corresponding $\Delta x = b$ meters and

$\Delta(cB) = 1/(cB)$.

Then $b \geq (h\text{-}bar)^{1/2}$.

Further defining b-minimum as the minimum Δx and using the positive root in this analysis, then

b(min) =1.027E-17 meters.

Subject to the further justification below, we assert:

$E = F\text{-sub-B} \times b$

Where $F\text{-sub-B} = F\text{-sub-G} =$ the gravitational force at the spatial location of event B.

And on the planet surface, $F\text{-sub-B} = ma = m \times 9.8$ meters-sec^-2.

Then $E / m = a \times b = 9.8b$ meters / $(cB)^2$ or

$E / m = 9.8b / (c/b)^2$ and

$E / m = 9.8 / 9 (10^{-16}) b^3$ J-kg^-1,

Or we can write the expression:

$E / m / b^3 = 1.089E\text{-}16$ J kg^-1 or

E-sub-B / m = 680 eV / kg for one cubic spatial boundary.

Justification for Spatial Dimension b

Using uncertainty and similar to the derivation above, one estimation using neurological sensory communication as an upper bound on $\Delta x = b$ (in one dimension) for spatial boundary (frame) size is suggested by:

(Spatial Frame Width)$^2 \leq$ (h - bar) x (c) x (Time Required for Sensory Continuity)

Using orders of magnitude 10^{-34} J-sec (and adjusting for units) from wave mechanics and estimating the time required for sensory communication in the range 10^{-3} sec - 10^{-6} sec from synapse switching (potential change) rate, we would then estimate the magnitude:

$\Delta x = b \sim 10^{-14}$ to 10^{-16} meters (for example as an upper spatial bound) in order to perceive continuity from actual contiguity of frames.

The neurological bound approximates the largest frame or spatial size that could be perceived as the continuity of time and accommodates the calculated boundary dimension $\Delta x = b \sim 10^{-17}$ meters.

b(min) = 1.027E-17 meters was derived assuming t = cB so that c is assumed to be the maximum achievable velocity[iii] and as such defines a maximum sequential rate and a minimum allowable b.

Justification for Spatial Barrier Energy

Using a one dimensional example,

E = F x distance.

We are using the transformation t = cB where B = 1 / b and b has the spatial dimension of meters.

The energy associated with the distance b is a function of a force F acting upon a mass m at a particular set of spatial coordinates.

It follows, the innate force acting on the mass m in space is the gravitational force.

There are no external forces to be considered for the mass m for our purpose regarding the transformation associating time and space.

Verification of Calculations

E-sub-B is then a function of gravitational force.

Using the calculations above and for a single event B = 1,

we can also write, for the planet surface as an example:

$E / m = a \times b = b \times 9.8 \, (c / b)^{-2} = 9.8 / c^2 / b.$

Then $1.089E\text{-}16 = 9.8 / c^2 / b.$

And for the planet surface E-sub-B, we verify our unit of measure calculations:

b meters = (1 event / B events meter^{-1}) = 1.000 as a confirmation of the energy calculation 680eV / kg.

Independently, we can re-calculate the value of b using F-sub-G on the planet surface:

b = E-sub-B / m x (a)$^{-1}$ = 1.089E-16 / 9.8 = 1.111E-17 meters.

This should be the universal value of b and is independent of F-sub-G since the accelerations "g" cancel for any spatial position.

This value is larger than the allowed minimum calculated b(min) = 1.027E-17 by the difference 8.4E-19 meters and we find the calculated surface value to be approximately 8% larger than the minimum allowed value b(min) using the Earth gravitational acceleration a = g = 9.8 m-s^{-2} and using no transformations in this calculation. We do not pursue further calculations in the present scope. **(See Appendices for calculations.)**

Energy Change as a Function of F-sub-G = F-sub-B

Assuming mass m and boundary b are unchanged, then E-sub-B changes as a function of F-sub-G = F-sub-B the gravitational force at the location of physical event B.

This follows directly from

E-sub-B = F x b.

A smaller gravitational force leads to a smaller E-sub-B relative to the planet surface.

With different E-sub-B, clocks should appear to run at different rates in regions of higher or lower F-sub-G relative to the planet surface.

A fictitious force, like the Coriolis force or the weightlessness of orbit, should not affect the real force F-sub-G = F-sub-B.

Conclusions

Continuous time can be represented by a contiguous spatial sequence of frames, or boundaries $\Delta x = b$, while conforming to existing physics in the macroscopic sense and with our sensory perceptions.

The transformation $t = cB$ leads to the spatial frame dimension $b(min) = 1.027E-17$ meters and corresponds to 3×10^8 physical events in one second of time t.

For this model,

The surface barrier energy E sub-B per unit mass = 680 eV / kg has been defined.

E-sub-B is suggested to be a function of gravitational force and so a function of spatial location.

Perceptions of Earth-time and clocks are expected to experience different rates in regions of lower or higher gravitational force relative to the planet surface.

XIV ii.

Appendix A (Reference)

The difference in value between b(min) derived from the Schrödinger equation and b calculated empirically from Earth-surface F-sub-G is 8.4E-19 meters and proves to be the mean factor 1.079 or 7.9%.

The concept of continuous time t leads to exponential growth-decay:

$a = a\text{-sub-}0 \times e^{\wedge}$ (rate x time) where $e = \lim (n \rightarrow \infty)$ $(1 + 1/n)^{\wedge}n = 2.718$.

Applying the expression for time itself, then

$T(\text{new}) / T(\text{old}) = e^{\wedge}(r \times t) = e^{\wedge}(0) = 1$.

For time t itself, the rate $r = 0$ and there is no change in continuous time t so that one "second" of "time t" does not change. Time t is absolute.

Differential equations for centuries, e.g. the Schrödinger equation, assume time t is a real and continuous variable.

In the transformation $t = cB$, we need to treat continuity of time as a slight-contiguity of space.

In that case, we find 3×10^8 met/sec to be a large enough frequency (number n) to continue using the calculated value of $e = 2.718 = \lim$ as $n \to$ infinity of $(1+1/n)^n$.

But in a directional spatial sequential model, then space itself advances or grows per some rate different from $r = 0$.

As a one dimensional chalk line curves in a two dimensional blackboard, and as a two-dimensional earth-surface curves in three-dimensional space, then a change in 3-dimensional space needs to take place in a mathematical dimension higher than 3.

Postulating the higher number of dimension (vertices) to be 5 as in the Fibonacci infinite sequence, and considering physical events $B = 1 / b = \sec^2$, then we calculate the following for one second of time t:

$$(\text{V-sub-S} / \text{V-sub-0})^{1/5} = e^{(r_v \times t)^{1/5}} = e^{(.618^{(5-3)})^{1/5}} = 1.079$$

or a 7.9% decrease in physical events B (increase in one-dimensional size b) from the continuous-time model used to calculate b(min).

XIV iii.

Appendix B (Reference)

Another calculation for difference
b-empirical - b(min) = 7.9%:

In two dimensional physics,

$F = ma = m \times$ met-sec^{-2} becomes $F_2 = m \times a_2$

$= m \times$ met - sec$^{-3/2}$

and one physical event B would no longer have units of sec^2; instead,

sec$^{3/2}$.

The uncertainty principle then has a transformed h-bar, and now

$b(min) = $ (h-bar)$^{1/2} \times c^{1/2} = 1.779E\text{-}13$ meters.

Similarly, $E_B / m = 9.8 / c^{3/2}$ J-kg^{-1} per square boundary

$= 1.886E\text{-}12$ J-kg^{-1} per boundary and

b = E-sub-B / m / 9.8 = 1.924E-13 meters.

Then we again have the mean factor

= 1.079 or 7.9%

between b(min) and b (from F-sub-G) in two-dimensional space exactly the same as in three-dimensional space.

XIV iv.

Appendix C (Reference)

A General Fibonacci Calculation:

The Fibonacci infinite sequence was referenced in Appendix A,

$F(n) = F(n-1) + F(n-2)$ with seed values $F(0) = 0$ and $F(1) = 1$.

Ratios converge, and

$\lim(n\to$ infinity$)$ $F(n+1) / F(n) = \varphi = (1 + 5^{\wedge}1/2) / 2 = .618 \ldots$ and

$\lim(n\to$ infinity$)$ $F(n-2) / F(n) = \gamma = .382 \ldots$ and so on.

Writing an example expression for spatial dimension ≥ 3 per Appendix A

$$\iiiiiiii dV = \iiiii dV(0) \times \exp(r_v \times t)$$

where $r_v = $ r-sub-V $= \varphi^{\wedge}(D(n+1) - D(n))$

then

$dx/dx(0) = \exp(\varphi \wedge (D(n+1) - D(n))) \wedge 1/(n+1)$.

Except we are now doing math in another dimension, and while $e = 2.718$ in three dimensions, the base of natural logarithms should change in higher or lower dimensional space.

For example, in the case of 5 dimensions: $e \rightarrow e^{\wedge}(1/\gamma)^{\wedge}(5/2)$.

We quickly find $dx/dx(0) = 1.08$.

A different example, for the case of spatial dimension < 3:

The base e must change as a function of the power of B, i.e. in three dimensional space $B \sim \sec^{\wedge}2$ while in two-dimensional space $B \sim \sec^{\wedge}3/2$.

The difference in power of physical events B

$2 - 3/2 = 1/2$ and the two-dimensional $e = 2.718^{\wedge}1/2$.

Then we quickly find $dx/dx(0) = 1.08$ similar to the previous mean calculations for the difference between b(min) and b-empirical.

The (Fibonacci) calculations hold true for any spatial dimension n moving through n+1 with a dimensional adjustment for e.

Appendix S (Reference)

Derivative Coordinates for Two Added Vertices

A derivative, e.g. df(x)/dx is tangent to the curve f(x) similar to the 5-dimensional coordinates t_x and t_y from Appendix Q.

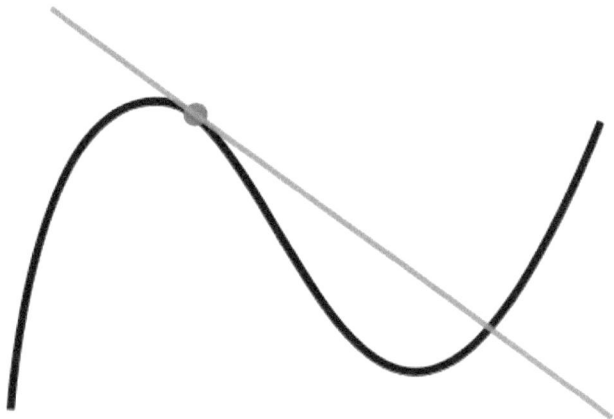

The derivative itself is independent from (orthogonal to) the point x, f(x).

The derivative is a function at the point x, f(x) relative to the points following and preceding the point x, f(x). It indicates the direction from the existing position to the next position.

In the Fibonacci spatial progression, we are traveling from the spatial dimension D=3 to D=5 and from the power-of-physical-events 2 to 3 as in the accelerations from 3 and 5 dimensional forces respectively.

The $\lim (n \to \text{infinity}) \, F(n-1)/F(n) = \varphi = 0.618$,

and this is the dimensional sequential growth rate. Then,

$$dt_x/dx_c = dt_y/dy_c = \varphi$$

and this matches the alternate view provided in Appendix Q. **The growth rates from Appendix C can now be more easily visualized.**

To understand the rate of spatial growth using dimension 3 traversing through dimension 5 as an example:

$$r_V = \varphi\text{^}(5\text{-}3) = \varphi\text{^}2 \text{ or } \varphi^2$$

meaning we have now added two extra dimensions (vertices) each having the growth rate φ so that the rate of additional volume growth is φ^2.

To understand the increasing value for 3-dimensional e = 2.718 (the base of natural logarithms) in higher dimensions D=5 and above, we need to review the dimensional geometry as follows:

Note: Here we apologize for confusion in nomenclature regarding the Fibonacci ratio and the Euler[3] angle both known as φ. Within the scope of this appendix, these will not be equated.

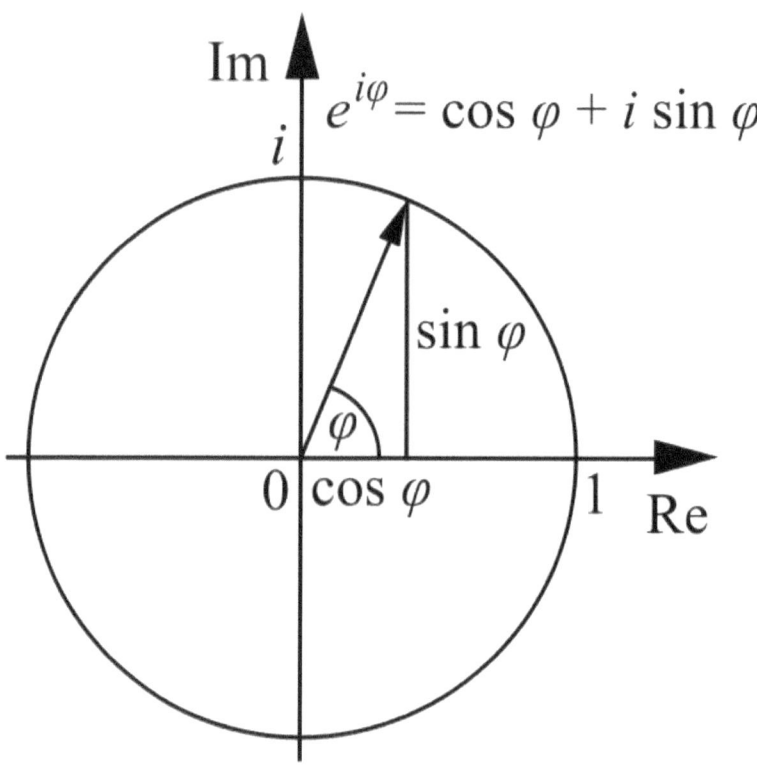

In a similar way to the rate of volume expansion and per Appendix C, the expansion of the logarithmic base (growth rate of the radius R in the above diagram) represents the addition of extra vertices. The number of vertices has grown from 3 to 3+2 = 5. Then,

$r_L = (1/\gamma)^{5/2}$

using D=5 traversing through D=8 as an example and where γ is defined in Appendix C:

$\gamma = \lim (n \to \text{infinity})\ F(n-2)\ /\ F(n).$

The physical concept

$i = (-1)^{\wedge}1/2\ \text{or}\ (-1)^{1/2}$

is then represented by the example vector:

$\mathbf{R} = (0,0,0,1,0).$

We can now physically extrapolate for charged points, lines, surfaces and volumes as well as their relative motions:

$\mathbf{E_3}$ is orthogonal to $\mathbf{B_2}$ where:

$\mathbf{E_3}$, $\mathbf{B_2}$, and $\mathbf{EB_5}$ represent open geometries moving within closed geometries of the same dimension, i.e. a radius within a circle and an open plane within a spherical surface respectively for $\mathbf{E}$ and $\mathbf{B}$.

We can express sequential values for E and B as below:

E: UP   0   DN   0

B: 0   DN   0   UP

And the corresponding vector sequence:

$$EB = EM = (0,0,1,0,0)\ (0,0,0,1,0)\ (0,0,-1,0,0)\ (0,0,0,0,1)$$

where the rate of vector progression through higher dimensional space is 1/4 cycle within b/4 meter and one complete cycle within the spatial progression b meters.

This is the nature of light.

The following diagrams represent achievable technology from the outline in appendices P through S.

These diagrams represent, among other things, the 5-dimensional computer model. The example atomic nucleus is Helium due to its multi-charge simplicity.

The smallest three dimensional energy ratio E / E_B is more than enough to provide spin state motion for the positive charges.

For example, one so-called electron-spin-split energy state transition (fine state transition $2_{p3/2}$ to $2_{p1/2}$) for Hydrogen easily provides for the real nuclear "spin" difference having the value $E = 4.5E-5$ eV with the wavelength 2.7 cm.

Energies for charge motion have already been achieved through techniques such as magnetic resonance[4].

XIV vi.

APPENDIX X

POSSIBLE KERNEL

In mathematics, a kernel is the smallest constituent or basis upon which everything else is built or advances.

For the low energy communication of hydrocarbons (perhaps the basis for the growth of life itself,) a suggested kernel is CH_3:

At once, the operational group is geometrically fixed, capable of exchanging hydrogen spin state energies within itself and among other operational kernels, and is prolific.

Related diagrams follow below.

Initialization of energies and continuity of the unaltered low energy wave within any actual kernal or kernel group should be critical to maintain natural sequences. As an example, radioactive charge energies are known to cause morphed growth sequences.

$$R-CH_3$$

$$R-\underset{\displaystyle H}{\overset{\displaystyle H}{\underset{|}{\overset{|}{C}}}}-H$$

XIV vii.

Appendix Z (Reference)

Further Examining Spin and E = hv

From $t = cB$, we can derive the following per the main manuscript:

$v = c / \lambda$ cycles per second.

For the H 2p3/2 - H 2p1/2 spin split transition, $\lambda = .027$ meters, then

$v = c / \lambda = b$ E+27 cycles per second, or

$v = b$ E+27 / (cB) cycles per event. It follows,

$v = (b^2 / c)$ E+27 cycles for the spin split transition, or

$v = (b^2 / c) (c^3 / \lambda)$ cycles, and

$v = (cb)^2 / \lambda$.

Equivalently, we can write:

$v = c^2 / \lambda B^2$. This has the form of a quantum expression.

We can define the maximum possible v,

$$v = c^2 / \lambda.$$

We can quickly show:

$$v = c^2/ \lambda = (b_3 c^0 \kappa^6)^{-1} = (b_5 c^1 \kappa^6)^{-1} = (b_8 c^3 \kappa^6)^{-1}. \text{ Then}$$

$n = \lambda v \kappa^6$ where $\lambda = nb3$ per Appendix E.

In this case, $\kappa = \gamma' / \varphi'$ and represents the 3-to-5 dimensional spatial transition per appendices D, E, and W.

Per Appendix Q, a one-dimensional radius in 5-dimensional space obeys a 6th root law.

We do not intend to generalize the spin split calculation and expression for higher dimensional v (meter^{-1}) in the present scope.

Closed Spatial Travel:

Consider the mass-matter that has the smallest possible natural (not externally excited) |energy| 2-D surface within 3-dimensional space.

This should be the surface having the radius r', where r' represents the 2-D cross-section radius from the single proton nuclear center of an H atom to the ground electron state for the same H atom.

Next, consider the spatial traverse for the single 3-D H (hydrogen) atom through 5-dimensional space:

From the 5-dimensional view, 3-D geometry is closed (curvature=1) in 5-dimensions. Then the 3-dimensional space must traverse a closed 5-D "circle" through the additional two (of the total five) dimensions in 5-D.

Except a 5-D "circle" is not the same as a 2-D circle that can be simply integrated into a closed surface and subsequently the surface into a "volume." While straight-forward, the next volume (5-D) integration involves two dimensions expanding into five.

In 2-D, the relationship for a closed-distance is $d_{c2} = 2\pi^1 r$ because the spatial change is only to-and-from a delta of one dimension. It directly follows, in 5-D the relationship for a closed-distance is $d_{c5} = 2\pi^3 r$.

The minimum allowed distance for the two dimensional surface to progress along the direction (0,0,"v/v") to "return to where it started" or to return "back onto itself" (close itself) is cb_3.

The increase in one-dimensional size (r) from 3 to 5 dimensions is .08^(5/3) per Appendix C.

Then,

$$2\pi^3 r(1+.08^{5/3}) = cb_3 \text{ where}$$

c = 2.998E+08 events and
$b_3 = 1.111E\text{-}17 \text{ meter-event}^{-1}.$

Then,

$$r = cb_3 \,/\, 2\pi^3(1+.08^{5/3}) = 5.29E\text{-}11 \text{ meters} = r_B. \text{ In other}$$
words,

$$r = r' = r_B \text{ the exact Bohr}^4 \text{ radius.}$$

To generalize:

For multiple mass-pair (m_p, m_E),

$$r = cn^2 b_3 \,/\, 2\pi^3,$$

and as a function of energy,

$$r = cn\lambda \,/\, 2\pi^3,$$

where r is the function $r = r(E/E_B)$.

To continue with the H example:

In 3D, the radius r is not fixed and the 3D geometry is not closed.

In 5D, while $2\pi^3 r_5 = cb_3$ and $s'/s = (1+.08^{5/3})\pi^3$,

in 3D, $s'/s = \pi/\varphi$ or $ds'/ds = \pi/\varphi$.

Arguably, for H:

There is normally no neutron, and the 3-to-5 (requiring a 5-to-8) traverse is not present.

In that case, H can be subject to the small-value G since K is 0 (zero.)

Then H, and H alone, may become a victim of the 3D (G) gravitational energy requirement $= E_B = 680eV/kg$ on the planet surface.

Arguably, were we constructed from 100% heavy-H (as in heavy-water where H is the single one-neutron H isotope,) then the hydrocarbons within us would not "age" because G would not apply in the presence of the stronger force K.

The three-to-five (similar in principle to the 5 to 8) traverse should be an electron transition. These shall be beyond the scope of the present text.

XIV viii.

A Technical Summary of *Changing Your Mind*

Per the main text and manuscript, we assert the transformation $t = cB$

where t (sec) = c (meter/sec) x B (sec^2 / meter) = c / b (event/meter) and where a single event = one sec^2 and where we need to remove "time" t from our units of measure for non-macroscopic physics.

It has been shown that

$b = 1.111E\text{-}17$ (meters), and that

$E_B = 680eV$ / $1kg$ / (b^3 met^3).

The Fibonacci infinite sequence can be described as follows:

$\text{Lim } (n\rightarrow\text{infinity}) \ F(n\text{-}1) / F(n) = \varphi = 0.618 \ldots$ and

$\text{Lim } (n\rightarrow\text{infinity}) \ F(n\text{-}2) / F(n) = \gamma = 0.382 \ldots .$

The growth rate of space itself for three dimensions expanding through five dimensions is

$r_V = \varphi^\wedge(5\text{-}3)$ so that

$V_5 / V_3 = e^\wedge(r_V)$ for one "second" of time t.

Similarly, the growth rate of the natural logarithmic base is

$r_L = (1/\gamma)^\wedge(5/2)$ so that

$e \rightarrow e^\wedge(r_L)$ for D=5 traversing through D=8.

Appendices A and C are attached to this text for reference.

Quantum mechanics is then described as $E / E_B = \kappa c / n$ and

$h = fn(E_B) = bE_B\kappa.$

The definition of quantum mechanics becomes mass x volume and defines the chemistry of the periodic chart of elements, and the Schrödinger[2] solution is replaced by

$E' = E_B(1 - p/n^2).$

Dimensional spatial curvatures are defined as $0 \le c \le 1$ with the terms

$0 = $ open, and

1 = closed.

Appendix K is attached to this text as a reference for curvatures.

Appendix L is attached as references for 5-dimensional geometry and 5-dimensional nuclear force respectively.

XIV ix.

Technical Summary of *Fifth Dimension*

Then there are two boundary conditions defining the natural elements:

1. E / E_B = mass x volume = atomic mass x 4/3 π r_A^3

2. $b / Q_0 = r_N$ x nuclear charge / nuclear mass

Diagram 6 is the best overall summary of *Fifth Dimension*.

~Inserted Note: This diagram is the same as Diagram 3 in the Diagram Section XI of the present text.~

Appendix S is attached as reference for the added two derivative coordinates.

Appendix Q is not attached except for Diagram 18.

Some diagrams from the original Appendices Q and S are included in the Diagram Section.

XIV x.

Cold Fusion Technical Summary

1. The charge attributes + and - have been shown to be dimensional attributes of mass.
2. The semi-attribute "electron spin" has been shown to represent proton spatial position within the five-dimensional nuclear interior.
3. Quantum mechanics and the dimensional nuclear surface force F_N independently derive a universal gravitational constant.
4. The uniform gravitational constant $K(D) = K(\Delta D) + G$ where $K(\Delta D) = 7.57E+28$ (Nwt kg^{-2} m ^{2}) for $\Delta D=1$.
5. $\kappa = \gamma' / \varphi'$ defines $h = h(E_B)$ and the transitional value $n = \kappa c$ as well as the transition-time expression $t = (1 + \kappa) / c$.
6. Centers of mass appear to be origins of coordinates for the sequential spatial traversal.
7. The dimensional model t=cB is proved by Appendix W.
8. The five dimensional computer model has further been defined by the material-related value λ.
9. The dimensional cold fusion model has been shown. A figurative example has been suggested.
10. A kernel (CH_3) for the hydrocarbon low energy communication has been suggested.
11. Facts of dimensional neutron mass have been shown in Appendix Q and have been suggested in Appendix Y.
12. The correct nuclear radius values for He and H have been shown.
13. The value ν as in $E = h\nu$ has been redefined and re-unitized in the absence of time t.

14. We now more clearly see the meaning of $E / E_B =$ mass x volume.
15. We now more clearly see the meaning of traversal events B where one second of time t is a representation of c events B.
16. We now understand the perception of time-dilation.
17. We now understand that to move forward we must first turn backward and review Diagram 21.

XIV xi.

Summary of *The Nature of Mass*

The Nature of Mass

Learning from the manuscript "A Mathematical Transformation of Variables . . ." (attached,) we now better understand the nature of mass.

Mass is the vehicle by which space is pulled through the natural spatial sequence.

The universal mass force is represented by the constant K(D) per Appendix W. A summary of the text *Cold Fusion* is attached as a reference.

We have seen throughout the Manuscript that the value of mass is not a function of spatial dimension. As a two-dimensional example, see Appendix B (attached.)

We also know that mass is two-dimensional in 8-dimensional space, one-dimensional in 5-dimensional space, and zero-dimensional (a point mass) in three-dimensional space. As an example, see Appendix Q (*Fifth Dimension, The Light to See.*)

We have calculated the hydrogen (H) nuclear (proton) radius in Appendix W from the zero-neutron unstable He structure:

$$r_{NH} = r_{NHe} / 4 = 7.7E\text{-}15 \text{ meter.}$$

There are no "points" in real three-dimensional space:

$$r_E = (3 / 4\pi)^{1/3} \times b_3 = 6.89E\text{-}18 \text{ meter.}$$

The *ratio* of "three-dimensional proton radius" / "three-dimensional electron radius" represents the mass force K(D) "pulling" 3-dimensional mass through 5-dimensional space per the natural sequence. Calculating exactly as in Appendix Z (*Cold Fusion*, attached) we now have the mass requirement:

$$m_P / m_E = (r_{P3} / r_{E3}) / \varphi = r_{NH}(1 + .08^{5/3}) / r_E / \varphi \text{ where } \varphi = .618 \ldots \text{ or,}$$

$$m_P / m_E = 1,836.$$

Note:

The arguable empirical value for the ratio is 1,833 and is within the factor 0.16% (0.0016) of the calculated value. The interchange of AMU and m_P "conventional" mass values in some calculations is a likely cause for small error. As in Appendix C, we do not intend to calculate the 3rd decimal for the value 0.08 in the present context.

The Nature of Higher Dimensional Mass

Similarly, for 5 and 8 dimensional mass (protons and neutrons,) the required curvatures of space and the universal mass force represented by K(D) lead to the following derivation:

As in Appendix Y (*Neutron Mass* attached,) the atomic nucleus is a three dimensional intersection between five and eight dimensional space. All nuclear mass is 3, 5, and 8 dimensional at the same "time."

All nuclear mass should "tip" a three-dimensional scale.

The 8-dimensional surface (neutron) mass pulling on the 5-dimensional linear (proton) mass also must pull on the entire 3-dimensional intersection mass. Atomically,

$$m_N / (m_N + m_P) = n_N / (n_N + n_P),$$

and this defines the curvature C from 5-to-8 and 3-to-8 dimensional space for the nuclear intersection, i.e.

$$(\varphi + \gamma) / 2 \leq C < \varphi.$$

(Diagram "INTERSECTION" should help to visualize.) Then,

$$0.500 \leq n_N / (n_N + n_P) < 0.618$$

and this agrees with Appendix Q (*Fifth Dimension*) and Appendix R (*Periodic Chart of Elements*, attached.)

Note: Standard H (hydrogen having no neutron) uniquely has curvature 0 per the attached Appendix Z. This is a special case.

Per Appendix Y, some curvatures within the above range are not allowed, and some energy states are not "100% present" exactly similar to electron "spin" energy states. The term "unstable" is inaccurate with respect to "time" but is accurate with respect to energy states.

Closed-Travel Energy

Closed spatial travel for three dimensions through five has the five-dimensional energy:

$$E_C = (Km_1m_2 / r_C^3) \times 2\pi^3 r_C.$$

For 5-dimensions pulling on 3,

$m_1 = n_P m_P$ and
$m_2 = n_E m_E.$

For 5-dimensions being pulled by 8,

$m_1 = n_P m_P$ and
$m_2 = n_N m_N.$

Then for the force pulling on *3-dimensional mass*, we can write:

$$E_C = (2\pi^3 K)\, n_P^2 m_P\, m_E\, /\, r_C^2,$$

and for the force pulling on *5-dimensional mass*, we can write:

$$E_C = (2\pi^3 K)\, n_P\, n_N\, m_P^2\, /\, r_C^2.$$

Inversely by r_C^2,

E_C for an atom is small, while E_C for a black hole is large.

Then for a 3-dimensional mass, $E_8 / E_5 = E_{CR}$ where

$$E_{CR} = (n_P\, n_N m_P^2\, /\, n_P^2 m_P\, m_E).\ \text{Then}$$

$$E_{CR} = (n_N\, /\, n_E)\, (m_N\, /\, m_E)\ \text{or}$$

$$\mathbf{E_{CR} = 1833\ x\ (n_N\, /\, n_E).}$$

Technical Conclusions
The Nature of Mass

1. Mass is proved to be a function of space.
2. Atomic nuclei are intersections between 5 and 8 dimensional space similar to black holes.

3. The Periodic Chart of Elements represents mass-space intersections that disappear for curvature $C = \varphi$.
4. Quantum mechanics is a (good) mathematical attempt to force-fit three dimensional thinking into higher dimensional space.
5. Particle physics is an extreme attempt to force-fit the universe into an ill-perceived three dimensional concept.
6. Hydrogen atoms (alone) and hydrocarbon molecules suffer the quantum energetic consequences of $E_{CR} < 1$.

XV. Reference

References

1. **Bernoulli, Jakob**. *Tractatus de Seriebus Infinitis*, ~1713.
2. **Ludwig Boltzmann**. *H-Theorem*, 1872.
3. William Thomson (Kelvin.) *On an Absolute Thermometric Scale*, 1848.
4. Johannes Rydberg. Lund University, 1888.
5. Alexander Boldyrev and Jack Simons. *Theoretical Search for Large Rydberg Molecules*; Journal of Chemical Physics, 11/01/1992.
6. Alasdair Wilkins. *The New Breed of Giant Molecules*; http://lo9.com, 08/02/2010.
7. Lamb, Willis E.; Retherford, Robert C. (1947). "Fine Structure of the Hydrogen Atom by a Microwave

Method". *Physical Review* **72** (3): 241-243. Bibcode: 1947PhRv...72..241L. doi:10.1103/PhysRev.72.241.

Reference Texts

1. *Changing Your Mind*, *t=cB*; Penguin Group, 2012.
2. *The Modern Health Guide*, Penguin Group, 2012.
3. *Fifth Dimension*, *The Light to See*; Penguin Group, 2012.
4. *Cold Fusion*, Penguin Group, 2012.
5. *The Nature of Mass*, Penguin Group, 2013.

www.ingramcontent.com/pod-product-compliance
Lightning Source LLC
Chambersburg PA
CBHW022019170526
45157CB00003B/1291